The Remarkable Railroad Passes of Otto Mears

1892 silver filigree pass No. 362, issued to C.W. Gibbs and his family, with its presentation card and imprinted leather carrying case. Gibbs was one of Mears' civil engineers. (John Gibbs Engstrom Collection)

The Remarkable Railroad Passes of Otto Mears

William K. Strong

SAN JUAN COUNTY
BOOK COMPANY

Silverton, Colorado
1988

First Printing by The Silverton Standard and The Miner
Silverton, Colorado

Second Printing by Preferred Printing & Graphics
Phoenix, Arizona

First Printing, 1988
Second Printing, 1996

ISBN 0-9608000-6-9

Contents

ADDITIONAL PASSES REPORTED SINCE FIRST PRINTING

TYPE	NAME ON PASS	NO.
1888	Not Issued	None
1888	Jim Carlisle (?)	A372
1889 A	K. G. Cooper	33
1889 A	Jim Carlisle	70
1889 A	Wm. B. Storey	91
1889 A	C. J. Harris	198
1889 A	M. A. Sanchez	219
1889 A	A. E. (L?) Drake	251
1889 A	W. R. Smith	353
1889 B	W. P. Knapp	634
1889 B	Not Issued	None
1890	A. _ Wingate	80
1890	Wm. B. Storey	93
1890	L. N. Wood	115
1890	Leland Stanford	189
1890	J. R. Watson	231
1890	Alfred B. Iles	332
1890	Daniel J. Carr	418
1890	Capt. L. E. Campbell	569
1892	K. G. Cooper	?
1892	Dennis Sullivan	81
1892	Wm. B. Storey	109
1892	Julian Ralph	219
1892	Charles Taylor	241
1892	J. F. Brown	297
1892	John F. Bell	348
1892	W. C. Van Horne	400
1892	G. W. Pratt	479
1892	H. G. Clark	558
1892	H. L. Fagin	594
1892	Mrs. Harry Lynne Fagin	596

MORE INFORMATION ON 1890 PASSES

Studying the 1890 pass with a magnifying glass shows at the lower left a parallelogram enclosing the letter P and at the lower right, "BEDD". Mr. Bedd probably was the designer of the die. The logo is that of the manufacturer, Diamond Palace. A similar logo appears on the back of the 1889 pass with the manufacturer's full name.

The Remarkable Railroad Passes

Any information about Otto Mears' unusual buckskin, silver and gold railroad passes for the Silverton Railroad and the Rio Grande Southern R.R. must begin with the background of this remarkable man.

Otto Mears was born in Russia of an English father and a Russian mother in 1840. After being orphaned at an early age, he lived with his maternal uncle's large family in Russia until going to live in England with his father's relatives for two years to the age of 11. Perhaps he did not get along well with his English relatives. They shipped him off to New York to live with yet another relative. He stayed a year there before being sent to San Francisco to live with an uncle. By the time he arrived, that uncle had left for Australia. The rest of his upbringing was by a woman who had befriended him on the ship in the long passage from New York (via horse over the Isthmus of Panama) to San Francisco.

Otto, at most 13 years old, earned his keep by working in stores, for newspapers, by selling ice cream, as a paper boy, and even working as a tinsmith. At the outbreak of the Civil War, he enlisted in the California Volunteers in August, 1861. Following three years of service in New Mexico, he was discharged at Las Cruces. After he worked for his first civilian employers in Santa Fe for six months, they financed him in starting a general store. The first was in Santa Fe, the next in Conejos, Colorado, and the third in Saguache, Colorado, about 1866. There he became a civic leader and involved in lumbering and wheat farming. This led to his becoming active in Colorado politics and in the transportation industry.

Otto Mears needed a dependable way to carry the wheat grown in the Saguache area to market in the

Arkansas Valley north of Saguache. Since there was none, he got a charter from the Territory of Colorado to build a wagon toll road over Poncha Pass to reach the Arkansas Valley. The Poncha Pass Wagon Road Company was financed with investors' capital, and its construction began in 1870.

The financial success of his first toll road soon led him to set up other toll road companies. The toll roads he founded grew to over 200 miles through the mountains of southwestern Colorado—from Saguache to Lake City, Gunnison, Ouray, Telluride and Silverton, and over Marshall Pass, Dallas Divide and Red Mountain Pass, among others. Those around Silverton and Ouray were nicknamed "The Rainbow Route." The name later was applied to the Silverton Railroad.

Mr. Mears became involved in several other businesses, including Colorado silver mines and mills, general stores, the Mack Brothers Motor Car Company (predecessor to Mack Trucks), and at least six railroads. One was the Chesapeake Beach Railway Company, which ran from Washington, D.C. to Chesapeake Beach, Maryland. He also found time in his early years to serve as a negotiator with Indian tribes. Mr. Mears was a king-maker in Colorado politics for many years. He lived to 91, passing away in Pasadena, California on June 24, 1931.

More information on the life of this unique man in Colorado history is found in *Otto Mears: Paradoxical Pathfinder* by Michael Kaplan, San Juan County Book Company, Silverton, Colorado 1982.

Lines Out of Silverton

In 1887, at 47 years of age, Otto Mears founded the Silverton Railroad to serve the mines north of Silverton at Chattanooga, the Red Mountain area and Ironton Park. By the time deep snow halted initial construction in January, 1888, five miles of three-foot narrow gauge track had been built north from Silverton. After snowmelt in 1888 the new railroad was extended into Red Mountain

Town and Ironton. In 1889 the Silverton was completed by building tracks into Albany, the unique turntable in Corkscrew Gulch, and a curved trestle across Red Mountain Creek, all on its main line. At most the Silverton Railroad was about 25 miles long. It operated through 1920, with formal abandonment in 1922.

Silverton had the distinction of being served by no less than four railroads, all narrow gauge. The Denver & Rio Grande completed its Silverton Branch from Durango in 1882. The line is still in operation by the Durango & Silverton Narrow Gauge Railroad. The Silverton Northern Railroad, also founded by Mr. Mears, was incorporated in 1885 to serve areas east and north of Silverton. Its tracks extended up the wide Baker's Park valley formed by the Las Animas River to Eureka and later up the narrow Animas Canyon to Animas Forks, elevation 11,300 feet. In many places its tracks were laid on one of the Mears' toll roads. The Silverton, Gladstone & Northerly Railroad was built in 1889 by the owners of the mill at Gladstone, but by 1915 it was owned by the Silverton Northern.

The Rio Grande Southern

Otto Mears began the Rio Grande Southern in 1890. Chief Engineer Charles W. Gibbs was assigned to survey its route from Ridgway on the Denver & Rio Grande tracks, south to Telluride via a spur, through Ophir Loop and Lizard Head Pass, Rico, Dolores, and finally into Durango where it again met the D&RG. Next Mr. Mears promoted investors to buy its stocks and bonds. In one trip to New York in late 1890 he raised $4,000,000. Track laying was completed by late 1891. The first train ran over its 162 mile length in January 1892.

At first the RGS was highly profitable from hauling silver ore out of the San Juan mountains and supplies and men in to work the mines. The repeal of the Sherman Silver Purchase Act and the resulting Silver Panic of 1893 so cut its business it was unable to pay its debts. Otto Mears lost control when the RGS went into receivership in

August 1893. The D&RG, the principal bondholder of the Rio Grande Southern, came into control. The RGS survived until 1952 hauling agricultural products, livestock, coal, lumber, ore and passengers.

1880s and 1890s Passes

Until 1906 railroads commonly issued free passes for passengers to travel on their trains. Passes were issued to employees, families of employees, favored shippers, clergymen, politicians, judges, and anyone else whose good will the management thought might further the interests of the railroad. Most passes were printed on banker's paper and by custom were good for a limited time, from one trip to sometimes as long as a year. It was thought by Congress that railroads might be buying influence with the passes. Their use was drastically curtailed by the Hepburn Act of 1906 so that future passes could be issued only to very limited classes of persons.

Otto Mears decided to do something out of the ordinary with some of his passes. Since his railroads were built to serve the transportation needs of western silver mines, what could be more appropriate and attention-getting than to make the passes of silver?

His first pass, though, for the Silverton Railroad for 1888, was made of buckskin, conjuring up ideas of the Wild West in the imaginations of the recipients. His next pass, for the Silverton Railroad for 1889, was made of stamped silver. In 1890 the Silverton's pass was in the form of a watch chain charm or fob. There was no silver pass for 1891.

For 1892, a silver filigree pass was issued good on both the Silverton and the Rio Grande Southern. Three of the 1892 passes were even made in gold! The Silver Panic of 1893 brought an end to wide distribution of the silver and gold passes. Three stamped silver 1893 prototype passes for the Silverton were made. A few silver filigree passes were produced for the Silverton for later years.

Information Sources

Much information on the passes has been lost in the fog of history. The late Josie Moore Crum of Durango wrote an article about the passes published in The Railway & Locomotive Historical Society Bulletin No. 73 issue for May, 1948. She identified 54 of Mears' non-paper passes she knew to exist. Morris W. Abbott, also deceased, became an expert on the Mears passes in building his extensive collection of them. He compiled a longer unpublished list of about 170 passes he knew to exist and a few more suspected he could not verify. Some of his "suspected" passes now have turned up. Robert W. Richardson and Jackson Thode also have had more passes reported to them. The author's research on the passes has revealed a few more.

The information on the Mears non-paper passes comes from the above sources. Many passes could not currently be located nor the names of their present owners verified. The data are based on the latest information available, and in case of variation between a list and information from a present owner of a pass, the latter is used.

Face of the 1892 silver filigree pass issued to C.W. Gibbs, who as chief engineer surveyed much of the Rio Grande Southern's lengthy route.

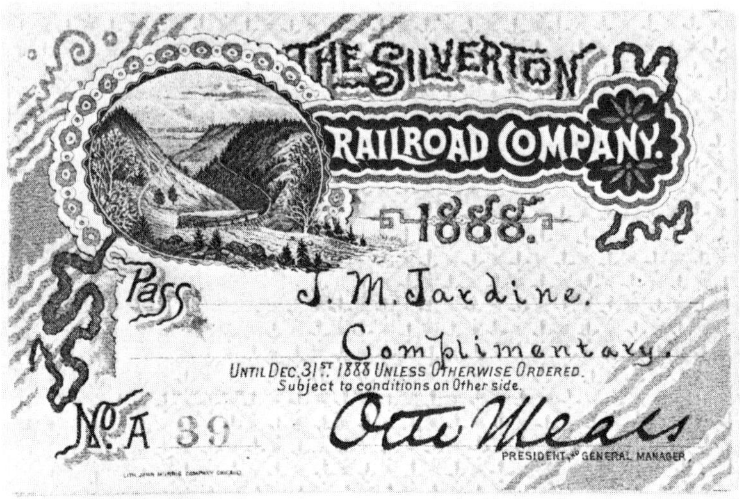

Face of 1888 buckskin pass No. A39, issued to J.M. Jardine for the Silverton Railroad Company. (Courtesy, Colorado Historical Society)

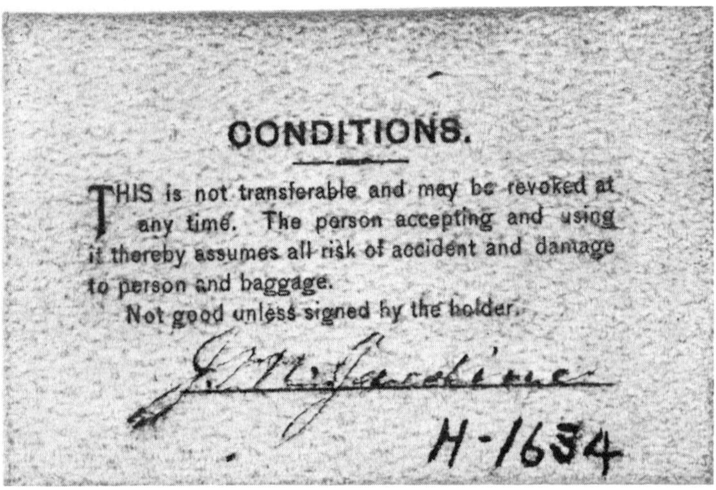

Reverse of 1888 buckskin pass, which required the signature of the holder in order to be valid for passage over the narrow gauge route. [H-1634 is a catalog number.] (Courtesy, Colorado Historical Society)

1888 Buckskin Passes

Probably fewer of the buckskin passes have survived than the later silver passes, due in part to deterioration of their organic material and in part to the lack of intrinsic value of buckskin.

The 1888 buckskin pass was about 2.5 by 4 inches in size, lithographed on light colored buckskin (perhaps cowhide) with light green and black printing. It shows a mountain railroad scene in an oval in the upper left, with green diagonal lines in the style of the paper passes of the era. The recipient's name was hand written. The digits, prefixed "A", were printed in red. Each appears to have been signed by Otto Mears. The passes were made by John Morris and Company of Chicago. No records are known to exist to show how many were made. The highest number on a known buckskin pass is A495.

Recipients of the 1888 buckskin passes included members of the Mears family, employees of the Silverton, and prominent people in Silverton and Durango. The most notable historical figure known to have received one was Horace A. W. Tabor, Leadville mine owner, politician and consort of Baby Doe.

Known 1888 Buckskin Passes:

PASS NO.	YEAR	NAME ON PASS
A 8	1888	R. E. Briggs
A 37	1888	Leopold Mayer and Family
A 39	1888	J. M. Jardine
A 42	1888	H. A. W. Tabor and Family
A 64	1888	Fred Walsen & Family
A107	1888	Geo. Ady
A127	1888	H. Silver & Family
157A	1888	Miss Edith Benedict
A208	1888	???
A290	1888	G. A. Scoville
A307	1888	A. P. Camp
A375	1888	John R. Hanna
A388	1888	Chauncey N. Nichols

A390	1888	Geo. V. Meserole
A439	1888	Hon. O. H. Abbott
A454	1888	Mrs. Otto Mears
A456	1888	Laura Mears
A458	1888	Cora Mears
A467	1888	C. F. Hoeckel
A484	1888	Lulu M. Becker
A489	1888	Dad Town
A495	1888	C. W. Gibbs

1889
Stamped Silver Passes

Probably late in 1888 Mr. Mears ordered his first silver pass, that for 1889, from Diamond Palace in Denver, whose name, city and logo appears on the back. Diamond Palace seems to have disappeared in the Silver Panic of 1893. This pass weighs an ounce and measures about 2.125 by 3.5 inches. Its upper left shows a mountain railroad scene, similar to that in the 1888 Buckskin, with a tiny steam locomotive pulling a string of four cars around a loop, perhaps freely representing the Chattanooga Loop on the Silverton. "H. Paul," probably the name of the engraver of the die, can be seen with a magnifying glass at the bottom of the scene. The year 1889 is stamped in relief, and a facsimile of Otto Mears' signature and "President" is carved into the lower right. The recipient's name and the pass number are engraved.

Morris W. Abbott, in his addendum to Mrs. Crum's article, identified a different version of the 1889 Silverton pass. He named the more common one shown opposite the Type A and the other, Type B.

The Type B has Otto Mears' signature and "President" raised from the surface instead of sunken into the pass as on the Type A, and only the first two digits of the year

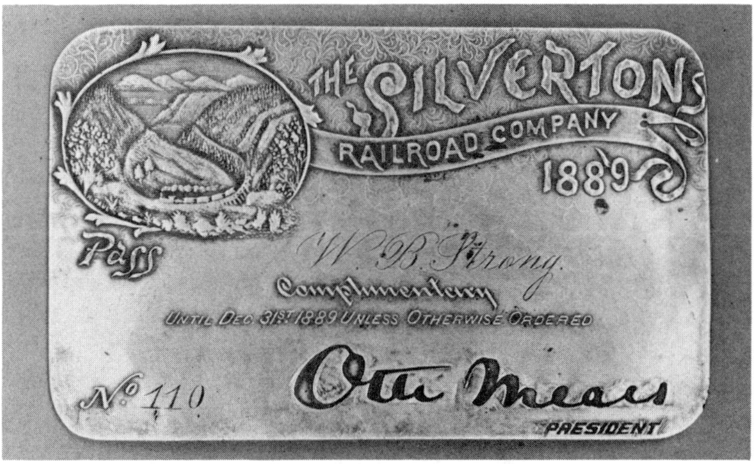

Face of 1889 Type A Silverton Railroad Company stamped silver pass No. 110 issued to W.B. Strong. Otto Mears' signature was recessed in the face of the metal on this particular style. (Author's Collection)

Reverse of 1889 Type A Silverton Railroad Company Pass. The logo reads: "Diamond Palace, Denver, Colorado." (Author's Collection)

''18'' appear on the pass (leaving it to the engraver to add the last two digits in the upper right and after ''Until Dec. 31st 18   ''). Diamond Palace's logo appears on the front below the railroad scene. The locomotive is reproduced in much more detail as a 4-4-0, and the wheels clearly are shown on the tender and cars where they were not on the Type A. The smoke from the locomotive is in a stream in the Type B where it is in puffs in the Type A. A close study of both shows a few more differences, such as in the tail of the ''S'' in ''Silverton'' and in the ribbons at the far right, among others.

Mr. Abbott supposed, and no one has questioned him, that the original Type A die was broken and that a different engraver made the die for the Type B. All known Type B passes have pass numbers over 500. The 1889 passes for Mr. Mears' daughters are Type B.

Known 1889 Stamped Silver Passes:

PASS NO.	YEAR	NAME ON PASS
???	1889	W. I. Gifford
???	1889	R. A. Lowe
007	1889A	A. S. Hughes
010	1889A	G. W. Kramer
046	1889A	I. A. Gothelf
047	1889A	L. Mayer
068	1889A	J. B. Grant
072	1889A	W. A. Bell
084	1889A	W. Clayton
087	1889A	Gov. Job A. Cooper
089	1889A	H. A. W. Tabor
090	1889A	William S. Jackson
105	1889A	J. W. Wingate
110	1889A	W. B. Strong
145	1889A	W. Smith
149	1889A	W. R. Marshall
152	1889A	William R. Adams
155	1889A	Casimero Barela
183	1889A	W. Ballinger
184	1889A	C. A. Bartholomew
189	1889A	J. W. McKinley

190	1889A	G. P. Copeland
196	1889A	H. H. Grafton
208	1889A	S. S. Merry
211	1889A	F. M. Murick
216	1889A	H. Prince
224	1889A	M. A. Vigil
228	1889A	R. M. Stevenson
256	1889A	F. A. Wadleigh
260	1889A	L. B. Eveland
267	1889A	Miss Edith Benedict
269	1889A	F. Brown
272	1889A	Moses Liverman
276	1889A	J. R. Hanna
282	1889A	A. M. Stevenson
283	1889A	Jay Gould
294	1889A	C. Tucker
296	1889A	J. J. Burns
297	1889A	S. K. Hooper
311	1889A	Frank O. Stead
313	1889A	R. R. Ripley
316	1889A	J. C. Helm
317	1889A	C. D. Hayt
326	1889A	H. M. Rhoads
333	1889A	C. A. Starr
339	1889A	Geo. Ady
348	1889A	G. H. Crosby
359	1889A	C. W. Gibbs
364	1889A	R. C. McNutt
375	1889A	E. L. Raymond
376	1889A	Fritz Theis
810	1889A	Carl R. Gray
000	1889A	Mr. C. R. Adams
514	1889B	W. S. Jackson, Jr.
519	1889B	Geo. R. Todd
527	1889B	Gov. H. B. Gillespie
550	1889B	J. M. Jardine
552	1889B	W. H. Crooke
553	1889B	Chauncey N. Nichols
557	1889B	L. U. Humphrey
562	1889B	C. S. Benson
563	1889B	John M. Kuykendall
564	1889B	Miss Laura Mears
565	1889B	Miss Cora Mears
569	1889B	C. F. Hoeckel

592	1889B	Fred Zell
610	1889B	J. F. Coykindall
618	1889B	Dr. John Elsner
620	1889B	Fred Walsen
???	1889B	Miss Anna M. Hayes

The 1889 Type B railroad pass issued to Otto Mears' daughter Cora and bearing No. 565. On the face of this style of pass, the railroad president's name was raised. (Courtesy, Colorado Historical Society)

1890 'Watch Fob' Passes

These small silver stamped passes were about 1.2 inches wide by 1.5 inches high. "Rainbow Route" appears in the top blue enamel band, a railroad track in forested mountains is shown in the center, and "Silverton Railroad" is in the bottom blue enamel band.

The back reads: Pass
(name of recipient)
Complimentary
Until December 31st 1890
Otto Mears
President
No. (serial number)

Some have a ring at the top to attach to a chain, but others do not.

Face of the 1890 "watch fob" or charm style of Silverton Railroad pass. The facial lettering on this type was silver on a blue enamel background.

Reverse of the 1890 fob or charm pass, issued as No. 141 to sleeping car magnate George Pullman. (Courtesy Paul P. Thompson from his collection)

Gold Lockets

At least two gold lockets were made from the watch fob die. The sides are identical and hinged at the top which has a loop. They have the blue enamel background for "Rainbow Route" and "Silverton Railroad." The one locket examined has a diamond on one side and a ruby on the other.

Inside the locket is engraved:

"Gold from
Sunnyside Extension Mine
Presented to
L.A. Stoiber
by Silverton R. R. Co.
Otto Mears"

The other gold locket was presented to Rasmus Hanson, but the misspelling of his last name as "Hansen" has gone down in history. Both Mrs. Stoiber and Mr. Hanson were owners of mines in the Silverton area. These do not say they are passes, so they are not listed with the passes. The "Hansen" locket now is in the Everett DeGolyer Library at Southern Methodist University in Dallas, Texas.

Known recipients of the 1890 passes:

PASS NO.	YEAR	NAME ON PASS
7	1890	J. J. Burns
28	1890	C. Tucker
29	1890	C. F. Hoeckel
35	1890	J. W. Deen
61	1890	Dr. Wm. A. Bell
64	1890	C. W. Gibbs

The face of Lena A. Stoiber's 1890 gold locket.

Inside of Lena A. Stoiber's unqiue 1890 Rainbow Route gold locket.

76	1890	Col. M. Breen
79	1890	J. W. Wingate
89	1890	Louis King
129	1890	Miss Edith Benedict
141	1890	Geo. M. Pullman
189	1890	Leland Stanford
221	1890	C. F. Meek
228	1890	John Adams
350	1890	C. A. Tripp
389	1890	George W. Seaman
409	1890	C. W. Bartholomew
418	1890	Daniel J. Carr
440	1890	James F. Gardner
453	1890	R. A. Lowe
461	1890	R. F. Crowell
470	1890	Sam W. Jones
475	1890	Fred Zell
488	1890	F. M. Duncan
494	1890	John J. Crooke
527	1890	H. G. Prosser
546	1890	F. F. Hogate
552	1890	Laura Mears
553	1890	Carl Gray
571	1890	J. F. Vaile
???	1890	Alfred B. Iles

1892 Filigree Passes

The most delicate of the Mears passes was the Rio Grande Southern - Silverton RR filigree pass for 1892. The silver pass was 1.5 inches high by 2.75 inches long (38 by 70 mm). To its solid silver border fine braided silver wires were attached to form a dense ornate web joining the border to a center plate. Light can be seen between the openings of the wires. Overlaid on the web were more designs in the form of fern leaves. Solid letters and numbers were added to read, at the top, "Rio Grande Southern RR"; at the bottom, "Silverton RR"; on the left,

Face of 1892 Rio Grande Southern-Silverton Railroad silver filigree pass No. 249 issued to George W. Childs and his family. The letters in the railroads' names were most likely individually made because of notable variations. (Courtesy of Paul P. Thompson from his collection)

Enlargement of part of the reverse of the Childs silver filigree pass of the 1892 series. (Courtesy of Paul P. Thompson from his collection)

"Pass"; and on the right, "1892". The letters appear to be hand made and applied by hand to the passes because of the irregularity of the letters and their placement on various passes.

The center plate, probably stamped, has ample space for hand engraving the recipient's name and the pass number following the letters "No", surrounded by a fine floral design. Otto Mears' reproduced signature appears raised on the plate in the lower third, and under his name in small letters is "Pres't." Small dots line the inside border of the plate.

Fine leather cases for the filigree passes were made imprinted:

COMPLIMENTS OF THE
RIO GRANDE SOUTHERN
— — — —and— — — —
SILVERTON RAILROADS

A card with the passes reads: "This pass is made of Colorado silver by native workmen."

Josie Moore Crum's 1948 article suggested that reports that over 500 filigree passes had been made was "pure sensationalism" and it was unlikely that more than 50 filigree passes ever existed. However, in her later book on the Rio Grande Southern she acknowledged that historians had found invoices in the records of the Rio Grande Southern for the filigree passes totalling nearly 600 in silver and three in gold. The invoices were from "S. Spitz, Manufacturer of Mexican Filigree Jewelry" on the South Side of the Plaza in Santa Fe. The invoice price of the silver passes was $4.00 each and of the gold, $15.33 each, plus the price of the gold itself.

Handmade filigree was an art form well-developed in Mexico and the Southwest in the late 1800s. While Mr. Spitz probably did not employ enough artisans in his shop to fill orders for that many filigree passes in a short time, perhaps he had many made by independent craftsmen.

The filigree gold pass is 1.4 by 2.5 inches (36.5 by 67

mm), a little smaller than the silver, and weighs one ounce. Mr. Mears' signature is larger than on the usual silver model. The "No" is larger and has no floral design around it. There are no dots near the inner edge of the center plate. The gold filigree passes also have a false interior border around the main filigree design connected to the outside edge by filigree loops in the manner children were taught penmanship many years ago.

The two known gold filigree passes were issued to "Miss Cora Mears" (No. 336), Otto's daughter, and to "A. J. Clark" (No. 266). The recipient of the third gold filigree pass is unknown. Morris Abbott reports that U.S. Senator H. M. Teller is said to have had one. Dave Day, a newspaperman close to Otto Mears for many years, also might have had the third. We will never know unless it appears some day. It was reported to Robert Richardson that the third gold pass was melted for the value of its gold.

One silver filigree, No. 366, issued to Mrs. A. Downes, is the same size as the gold filigree, and its center plate is similar to that of the gold, but Otto Mears' signature and the word "PRESIDENT" appear to be separately formed and soldered on. It does not have the false interior border found on the gold passes. Does this represent a Type B filigree pass, or is it unique?

The known silver and gold 1892 filigree passes are:

PASS NO.	YEAR	NAME ON PASS
7	1892	Carl L. Gray
10	1892	Mrs. A. S. Hughes and Family
11	1892	S. K. Hooper
16	1892	Mrs. J. J. Burns
17	1892	N. W. Sample
20	1892	T. H. Wigglesworth
21	1892	D. H. Moffat
29	1892	Walter S. Cheesman
30	1892	Mrs. E. R. Murphy
32	1892	Mrs. J. S. Hubbard
34	1892	Dr. J. W. O'Connor

35	1892	F. A. Wadleigh
36	1892	C. Tucker
41	1892	Mrs. Fred Walsen & Family
42	1892	Mrs. Moses Liverman
48	1892	Mr. & Mrs. J. F. Vaile
51	1892	Henry Collbran
57	1892	W. S. Jackson
70	1892	Charles Fargo
76	1892	J. B. Grant
86	1892	J. R. Hanna
97	1892	Chauncey Nichols
100	1892	Louis King
102	1892	W. B. Trufant
106	1892	A. P. Camp
110	1892	Geo. Fisher
117	1892	F. F. Struby
132	1892	George W. Holdrege
135	1892	Charles F. Mayer
138	1892	Timothy B. Blackstone
175	1892	John King
218	1892	A. T. Gunnell
229	1892	H. B. Adsit
233	1892	T. M. Patterson
249	1892	Geo. W. Childs and Family
269	1892	Lester G. Brewer
272	1892	M. J. Powers
274	1892	Joseph H. Maupin
288	1892	Victor A. Elliott
289	1892	Joseph C. Helm
290	1892	Charles D. Hayt
304	1892	John M. Kuykendall
306	1892	H. L. Kuykendall
309	1892	Frank Brown
360	1892	Richard Linthicum
362	1892	C. W. Gibbs
366	1892	Mrs. A. Downs
369	1892	Joseph P. Airy
374	1892	George W. Seaman
378	1892	Fred Walsen
393	1892	Mrs. Cooper Anderson
394	1892	Alex Anderson
399	1892	Mrs. Henry Cadwell
409	1892	James J. Ford
415	1892	Amos Walthers

421	1892	C. A. Tripp
422	1892	E. E. Hoffman
452	1892	J. W. Maynard
466	1892	E. B. Slosson
489	1892	Charles S. Thomas
493	1892	Chas. A. Bartholomew
528	1892	W. A. Bell
534	1892	Mrs. N. P. Hill & Family
541	1892	Jos. R. Stonebraker
550	1892	George Ady
555	1892	H. Wheeler
560	1892	Paul Morton
578	1892	Adelaide G. Stimpson
???	1892	Horace Clark
???	1892	John Burns
???	1892	Cole Lydon
266	1892G	A. J. Clark
336	1892G	Miss Cora Mears

Other Silver Passes

There are a handful of other Mears silver passes.

A large filigree pass for the RGS and Silverton, No. 1400, with an all-engraved center plate was issued for 1893 to Mr. Charles Tarbell. Another 1892-type regular-sized silver filigree, No. 605, was issued to J. N. Hannan (probably J.R. Hanna) for 1897. It has an all-engraved center plate without the cast floral pattern and without the "No.", substituting "STATIONS" for "PRESIDENT." Two 1892-type filigree passes bear the year 1893.

Three stamped silver passes were made for the Silverton Railroad for 1893. None were issued, although one is reported to have been engraved with the name "M. Liverman," the General Manager of the Silverton. These prototype passes combine elements of the three earlier silver passes. The size and method of manufacture resemble the 1889 stamped silver; the center design is from the 1890 watch chain charm or fob (without the blue enamel background for the lettering); and the center is surrounded by a fine raised pattern of lines after the 1892 filigree design.

A summary of known later passes:

PASS NO.	YEAR	NAME ON PASS
590	1893	Dr. & Mrs. E. J. Kinney
606	1893	Geo. W. Ristine
1400	1893	Mr. Charles Tarbell
???	1894	C. F. Hoeckel
605	1897	J.N. Hannan

Face of prototype stamped silver pass for the Silverton Railroad for the year 1893. Only three of these were made, and none were ever issued.

Numbers & Quantities

The lack of consecutive serial numbers on the known passes of any one year or on the non-paper passes in general has piqued the curiosity of students of the Mears passes. Mrs. Crum suggested that numbers on paper passes issued for the same year could have filled in the gaps. One story is that Mr. Mears assigned numbers at random, taking the last two or three digits from the weight of loaded cars shown on papers in his office.

Original RGS documents in the Southwestern Studies Center in the Fort Lewis College library in Durango provide valuable evidence of the quantity of 1892 filigree passes made. Three invoices from S. Spitz show billings for a total of 579 passes, including the three gold. RGS documents (vouchers and bills for vouchers) show that the RGS acknowledged its obligation to pay for a total of 594. The last RGS bill for voucher for the 1892 passes shows that it is for those numbered 590 through 598. The corresponding invoice from S. Spitz is dated March 29, 1893.

Because of the close correlation between the quantity of 1892 filigree passes shown in the business records of the RGS and the highest pass number shown in its records, it is highly probable that the filigree passes were numbered consecutively.

The only clue on whether the 1889 passes also were made in a quantity of over 500 and were mostly consecutively numbered is that about an equal number of the 1889 passes and of the 1892 passes have survived to be listed above. No business records are known to exist to give more information on the number of 1889 silver passes made.

In the 1940s it was thought that none of the non-paper passes bore identical numbers, but a few years ago two with the same numbers, but for different years, came to light. Then more with the same numbers appeared. Of the passes identified since 1948, three have No. 7, and twenty other pass numbers are duplicated on passes for different years.

Since no two numbers on known passes of the same year are identical, it is reasonable to believe that records were kept of pass numbers for any given year to avoid duplication.

Pass Recipients

The identity, status and accomplishments of the recipients of the passes is another interesting subject. Based on information compiled by Morris Abbott, mainly from Jackson Thode and updated by Jackson Thode as new names appeared, they fall into four general categories:

1. Government officials, including U.S. Senators, Colorado governors and members of the executive branch, Colorado legislators, mayors and judges;
2. Railroad officials of the Silverton, Rio Grande Southern, Denver & Rio Grande, and persons prominent in the larger roads and industries connected with railroading;
3. Colorado bankers, newspapermen and businessmen, particularly those who shipped large tonnages such as owners of mines and mills; and
4. Otto Mears' family members, old friends, investors and poker-playing buddies.

Of course, there are others who are not so easily categorized. For example, Josie Moore Crum tells that as a teenager, Mrs. Henry Cadwell was given an 1892 filigree pass in recognition for her nursing a man badly burned by steam when the RGS locomotive in which he was riding tipped over.

Names on (or reported to be on) the passes sometimes vary from the usual spelling of the recipient's name. For example, 1889 pass No. 46 was issued to "I. A. Gothelf". The usual spelling of his last name was "Gotthelf". The name variations undoubtedly are just human error, perhaps that of the person who wrote the order to the engraver, or of the engraver. The difficulty of the modern observer in reading the old English script may contribute. The spelling of recipient's name as it appears on the pass is used here.

The names and identities of some of those given the non-paper passes are below. The number after the person's name is the quantity of known passes issued to that person.

Hon. O. H. Abbott - Justice of the Peace
William R. Adams - Mayor of Alamosa 1885-86, Colorado State Senator
Mr. C. R. Adams - (Charles) Ute Indian Agent, Los Pinos
H. B. Adsit - (Harry) Cattleman, Norwood, Colorado area
Geo. Ady (3) - Railroad Official DSP&P, GP&TA, UP, GPA
Joseph P. Airy - Mayor, Red Mountain Town 1884, foundry owner
Alex Anderson - Auditor of Silverton Railroad
Casimero Barela - Colorado legislator, Trinidad & Las Animas
Chas. A. Bartholomew (3) - Possibly Ute Indian Agent
Dr. Wm. A. Bell (3) - Backer of Palmer's D&RG
C. S. Benson - Silverton businessman
Timothy B. Blackstone - President, Chicago & Alton Railroad
Col. M. Breen - Merchant, mine owner, Silverton & Ouray
Lester G. Brewer - Secretary to Colorado Governor, Vice President DNW&P
R. E. Briggs - Chief Engineer D&RG 1888-1899, later MP&P
Frank Brown (2) - Manager, Bordeleau's Hardware, Silverton
J. J. Burns (3) - D&RG Superintendent of Transportation and 1st Division
Mrs. Henry Cadwell - Lady who aided injured in RGS wreck
A. P. Camp (2) - (Alfred) Durango banker
Walter S.Cheesman - Denver banker
Colin A. Chisholm - Chief Clerk, D&RG Legal Dept. 1891
A. J. Clark - Original shareholder of Silverton Railroad
W. Clayton - (Geo. W.) Early Denver capitalist
Henry Collbran - Colorado Midland official
Gov. Job A. Cooper - Sixth Governor of Colorado, 1889-1891
J. F. Coykendall - (John) Secretary to Vice President, UPRR, 1887-1892
John J. Crooke - Smelter and mine owner
W. H. Crooke - (William) Silverton mine and business manager
B. F. Crowell - (Benjamin) Member, Colorado Board of Capital Managers
J. W. Deen - (James) Division Engineer, D&RG at Salida
F. M. Duncan - Chief Dispatcher, D&RG, Denver
Victor A. Elliott - Colorado Supreme Court Justice, 1889-1895
Dr. John Elsner - Early Denver doctor
L. B. Eveland - D&RG Passenger Agent, Kansas City, Missouri

Geo. Fisher - Mayor of Telluride, mine manager
James J. Ford - General Agent, Denver, LS&MS, NYC&HR
C. W. Gibbs (4) - Civil Engineer & Surveyor, Silverton Railroad
I. A. Gothelf - (Isaac Gotthelf) Early business partner of Mears
Jay Gould - Railroad magnate
J. B. Grant (2) - Third Governor of Colorado & Vice President, Denver National Bank
Carl R. Gray (3) - Senior Vice President SL-SF, later President, GNRR
John R. Hanna (4) - President, City National Bank of Denver
Charles D. Hayt (2) - Colorado Supreme Court Justice, 1889-1899
Joseph C. Helm (2) - Colorado Supreme Court Justice, 1879-92, 1907-09
Mrs. N. P. Hill & Family - Denver society leader, wife of Senator
C. F. Hoeckel (4) - Printer & stationer to D&RG
George W. Holdrege - CB&Q General Manager
S. K. Hooper (2) - D&RG General Passenger & Ticket Agent
Mrs. J. S. Hubbard - Wife of D&RG Superintendent Buildings & Bridges
A. S. Hughes (2) - General Traffic Manager, D&RG
William S. Jackson (2) - Banker, Director, President & Receiver, D&RG
W. S. Jackson, Jr. - Colorado Supreme Court Justice
Sam W. Jones - Colorado Attorney General, 1889-1891
Louis King (2) - President Ouray Stage & Omnibus Company

John King - Leadville miner and Denver coal merchant
G. W. Kramer - (George) Manager, D&RG Express, 1880-1893
John M. Kuykendall (2) - Denver Omnibus & Stage Co.
Richard Linthicum - Denver Educator & Teacher ?
Moses Liverman (2) - General Manager, Silverton Railroad
R. A. Lowe (2) - D&RG passenger conductor
Cole Lydon - D&RG Division Supt., Fourth Division
W. R. Marshall - D&RG official
Joseph H. Maupin - Attorney General of Colorado, 1891-1893
L. Mayer (2) - (Leopold) friend of Otto Mears
J. W. Maynard - UP ticket agent, Council Bluffs, Iowa
R. C. McNutt - (Ruben J.) Locator of Sunnyside mine, judge

Miss Cora Mears (3) - Otto Mears' daughter
Mrs. Otto Mears - Otto Mears' wife
Miss Laura Mears (3) - Otto Mears' daughter
C. F. **Meek** - (Channing) General Manager, DT&FW, Yule Marble
Geo. V. **Meserole** - Early freight and mail contractor
D. H. **Moffat** - Denver banker, mine owner, railroad official
Paul Morton - Official of CB&Q, AT&SF RR
Mrs. E. R. **Murphy** - Wife of D&RG General Auditor
Chauncey N. **Nichols** (3) - Manager, Beaumont Hotel, Ouray, 1889
Dr. J. W. **O'Connor** - Chief Surgeon, D&RG
T. M. **Patterson** - Owner, Rocky Mountain News, U.S. Senator, 1891-1907
H. G. **Prosser** - Silverton Mayor 1887 and businessman
Geo. M. **Pullman** - Sleeping car magnate
E. L. **Raymond** - (George L.) Editor of Durango Herald
H. M. **Rhoads** - Lt. Colonel and Assistant Adj. General
R. R. **Ripley** - Pueblo District Attorney
Geo. W. **Ristine** - President, Colorado Midland Railroad
N. W. **Sample** - Former Baldwin employee, D&RG Superintendent
G. A. **Scoville** - Durango resident
George W. **Seaman** (2) - Silverton newspaper editor, businessman
H. **Silver & Family** - Secretary-treasurer of RSA & LA Railroad
E. B. **Slosson** - D&RG General Agent, Chicago, 1891
Leland Stanford - Southern Pacific Railroad Founder
A. M. **Stevenson** (2) - D&RG Attorney at Glenwood Springs, 1887-89
Adelaide G. **Stimpson** - Became wife of Alexander Anderson
W. B. **Strong** - President, Santa Fe Railroad
F. F. **Struby** - (Frederick) Denver wholesale grocer
H. A. W. **Tabor** (2) - Leadville Mine Owner, politician
Mr. Charles Tarbell - Storekeeper at Saguache, friend of Mears
Fritz Theis - Early Denver musician
Charles S. **Thomas** - Law Partner of T. M. Patterson, Ninth Colorado Governor
Dad Town - (R. B.) Early owner of Yankee Boy mine
C. A. **Tripp** (2) - MoPac Traffic Manager

W. B. Trufant - Superintendent, Denver Union Station, UPRR
C. Tucker (3) - (Cromwell) D&RG Freight Claim Agent, Denver
J. F. Vaile (2) - (Joel) General Counsel, D&RG
M. A. Vigil - Colorado House of Representatives, 1888
F. A. Wadleigh (2) - General Passenger Agent, D&RG
Fred Walsen (5) - Treasurer, Silverton Railroad
Amos Walthers - Ouray banker
H. Wheeler - Incorporator, Silverton & Animas Forks Toll Road
T. H. Wigglesworth - Civil Engineer, locator on RGS & SN RR
J. W. Wingate (2) - Director and Secretary, Silverton RR 1887-1891
Fred Zell (2) - Friend of Otto Mears

Miscellaneous

A significant number of the passes now are owned by museums. The Colorado Historical Society in Denver has the most, 29, including all of the Mears family collection. The Everett DeGolyer Foundation collection at Southern Methodist University is reported to have about a dozen, Colorado College in Colorado Springs has three, the Denver Public Library has three, and the Union Pacific RR Museum has three. Large private collections have been assembled, and some have been broken up by piecemeal sales.

Were the Mears non-paper passes actually honored for free rides on the Silverton and RGS? The more notable recipients of the passes may never have been in Durango, Silverton, Ridgway or the other towns served by the Silverton and RGS. To them, the passes would be only fascinating conversation pieces. But other passes were issued to residents of Durango and Silverton and to management personnel of the two railroads. Therefore it is probable that the passes actually were good for free transportation for the year issued. The conditions on the back of the 1888 buckskin pass tend to confirm the conclusion.

Souvenir Spoons

An often-seen picture of the 1893 prototype stamped silver pass for the Silverton RR from the Morris W. Abbott collection shows the pass attached to a silver

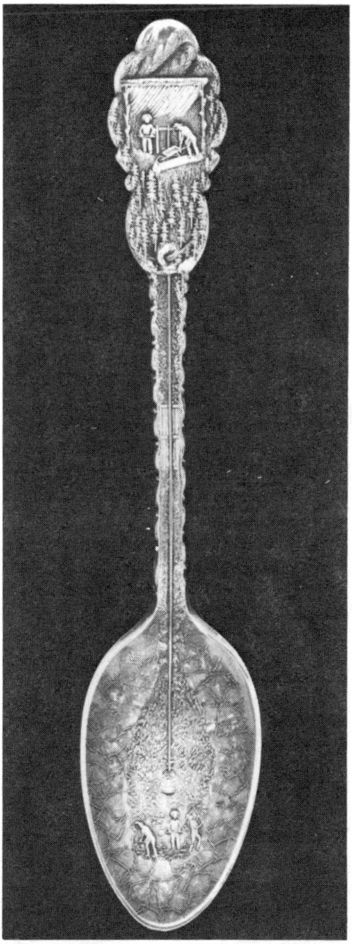

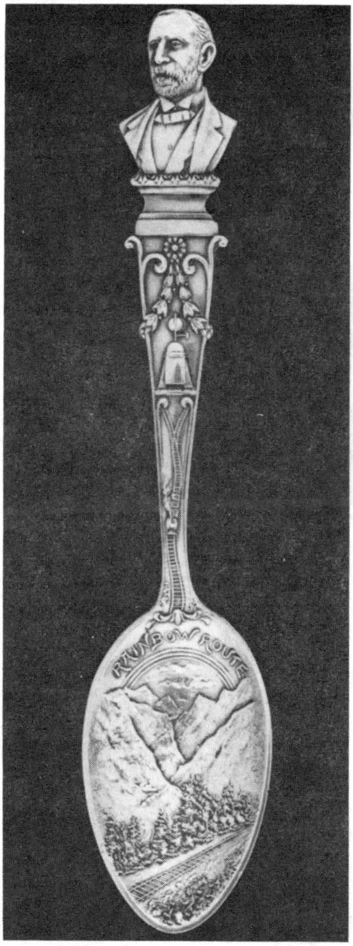

The souvenir spoon on the left depicts a mining scene, including a shaft running down the spoon handle. The Rainbow Route souvenir spoon on the right bears the likeness of Otto Mears. It is not dated. (Courtesy, Colorado Historical Society)

spoon with the likeness of Otto Mears at the top of the handle.

The souvenir spoon is reported to have been given to guests at a banquet in Silverton (Rainbow Route, P. 272). The Colorado Historical Society accession card for its two spoons states "Rico 1891". (Mr. Abbott identified two other spoons.)

The bowl of one spoon is designed similarly to the scene on the 1890 pass, showing mountains with tracks in the foreground, under "Rainbow Route". The second shows the top of a mine shaft on the handle, the shaft running down the handle, with miners at work in an underground mine on the back of the bowl.

About the Author

William K. Strong is a practicing attorney in Phoenix, Arizona. He became interested in the Otto Mears passes when he learned around 1980 that his family heirloom 1889 Type A pass was a rare item. He has had it since about 1940 and sometimes carried it with him as a child. His pass, No. 110, was issued to his great-grandfather, William Barstow Strong, General Manager or President of the Santa Fe from 1877 to 1889. By asking a lot of questions of friendly people in Silverton, reading several excellent books on Colorado narrow gauge railroads, making many telephone calls and writing even more letters, he was able to compile the data presented.

Special Thanks

An historical study of this magnitude does not just happen without the active help and co-operation of many other people. Morris W. Abbott, who passed away before I became interested in the Mears passes, deserves credit for most of the data presented here, which he assembled over a 20-year period. His pages of information were graciously supplied by his son, William Abbott. Jackson

Thode of Denver also was instrumental in encouraging me and providing information on several passes not listed by Mr. Abbott.

Special thanks go to the many people who answered inquiries. Responses to my inquiry letters varied from a polite "none of your business" to lengthy letters giving the person's recollections about the passes and even colored photos of their passes. Undoubtedly the publication of this work will cause more to come to light. I would be delighted to learn about any others for a future edition or supplement. Mallory Hope Ferrell also is recognized for his input and support. The home computer also may deserve some credit, for all of this was written on one, and data on the passes is stored on a hard disk, duly backed up!

The Holy Writ on Free Railroad Passes

The Beloit Daily Free Press - Beloit, Wisconsin
Cham Ingersoll, Publisher & Editor. Wednesday, December 27, 1899

The December number of "Locomotive Engineering", a valuable journal devoted to railway motive power and rolling stock, published the following incident in the railway life of our esteemed fellow-citizen, William B. Strong, ex-president of the great Atchison, Topeka & Santa Fe Railray System.

"Quite a long time ago a minister of the Gospel wrote Mr. William B. Strong, who was then a high official of the Michigan Central Railroad, requesting an annual pass over that system. The minister was duly supplied with a half-fare permit. This did not appease his desire, however, and he immediately instituted a more diligent pursuit of the coveted annual pass by writing Mr. Strong as follows:

'Dear Sir — You sent me a few days since a half-fare

permit, which, please, fully permit me to thank you for. As a proof that I should be given the annual pass I requested, I respectfully call your attention to the following passages of the Scripture:

"Exodus vi. 1 — 'with a STRONG hand thou shalt let them go.'

"II Chronicles xvi. 9 — 'strong in BEHALF of them.' (Half in this case means, not the half I now have, but the other half of the permit, so that I shall have a full free pass for the year.)

"I Kings ii. 2 — 'Be STRONG, and show thyself a man!"

"Jeremiah xv. 14 — 'I will cause thee to pass.'

"Ezekiel xxxvii. 2 — 'and he caused me to pass.'

"Joshua xxii. 19 — 'Then pass over.'

"If the above passages do not find or reach some responsive chord in your bosom, other language will, of course, utterly fail to impress you.'

"Mr. Strong replied as follows:

'It is an odd experience that the Scriptures can be made to sustain any doctrine or dogma if it be ingeniously applied. I thought I was complying strictly with the most liberal offers of transportation to be found in the Bible.

'Early in the history of the Jews we find the account of their immigration from Egypt; and certainly going in such numbers, they would be entitled to as low a rate of fare as could consistently be asked or granted by anyone, yet in the 13th verse of the 30th Chapter of Exodus I find the following: "This they shall give, everyone that PASSETH — A HALF." If this does not cover the case, I know not where to look for authorities.

'In one of your citations you refer to what you are pleased to call the "other half" of the permit which I sent you. I fear that it would be of little service, since our conductors, being better versed in mathematics than theology would be sure to collect full fare from anyone travelling on two half-fare permits.

'But to convince you that I look to Holy Writ as an authority for declining free passes, permit me to quote a few precepts on the subject of passenger transportation which I find in its pages, beginning with such as seem especially addressed to the passenger:

"I Kings xx. 39 — 'Thou shalt pay.'

"II Kings iv. 7 — 'Go and pay.'

"Ecclesiastes v. 4 — 'Defer not to pay.'

"Exodus xxi. 19 — 'He shall pay.'

"Jonah i. 3 — 'so he paid the fare and went.'

'In addition to these precepts to the passenger, I find the following injunctions to the railroad manager:

"Judges iii. 28 — 'Suffer not a man to pass.'

"Nahum i. 15 — 'The wicked shall no more pass.'

"Isaiah xxiv. 10 — 'None shall ever pass.'

"Jeremiah ix. 12 — 'that none passeth.'

'Perhaps after this array of Scripture authority I shall not be justified in sending you the annual pass desired, yet I find my sympathies stronger than my theology, and take pleasure in sending you herewith the pass requested."

Face of 1889 Type A pass No. 359 issued to C.W. Gibbs. The illustration on these passes is of a steam-powered locomotive pulling a train through mountainous terrain believed to be the artist-engraver's conception of the Chattanooga curve at the foot of Red Mountain Pass.

Index

K

L

M

N

O

P

R

S

T

U

V

W

Z